# Statistical Research with SAS:
# Design, Analysis,
# Interpretation

## Zhi Li

## July 2010

# Contents

# Histogram

## Problem Setup

Histogram is a graphic presentation that allows frequencies occurring in the data to be seen. There may be a class of 25 female students and 27 male students. And we can use a graph of 2 vertical bars to represent the frequencies. The bar with 25 units high represents the frequency of female students. Another bar with 27 units high represents male students. By comparing the length of these two bars, we can easily see how the difference of female and male students on their frequencies appeared in the class.

However, when we study continuous data, we will not have nice, easy to use categories such as we had in the students' case descriptions above. The case above has 2 categories - female and male - so we can count students into the categories by gender. For continuous data such as measurements like temperature, length of an object, or elapsed time, we need to define some categories to count the data frequencies.

For instance, the following SAS code presents a data set of annual sales value for a supermarket. The code defines an increment of 5 for the categories. From 20, each increment of 5 is a category. A category which is defined for continuous data is called a bin. Frequencies are counted by their bins. The percentage of total frequency for each bin is plotted as the bar tall. Certainly, sum of the bar tall equals to 1.

# SAS Code

```
data Supermarket;
input amount @@;
label amount = "Supermarket sales";

* Hypothetical Supermarket data;
datalines;
29.9  35.0
34.9  24.6
32.1  28.6
23.5  32.5
31.6  28.1
33.4  34.8
35.4  33.7
31.1  29.5
32.3  25.6
30.4  30.1
30.7  29.0
26.4  26.0
30.5  28.9
30.9  33.2
29.7  41.2
28.4  28.7
32.4  33.6
25.5  38.3
33.2  32.7
30.1  32.1;
run;
proc univariate data=Supermarket noprint;
     histogram amount;
     title 'Frequency Histogram for the
Supermarket Cash Sales in Thousands';
run;
title 'Frequency Histogram for the Supermarket Cash
Sales in Thousands';
proc univariate data=Supermarket noprint;
     histogram amount / midpercents name-'MyHist'
/* output frequency table */
     cfill = gray
     endpoints=0 to 45 by 5
     barwidth=5;
run;
```

# Interpreting the Result

**Frequency Histogram for the Supermarket Cash Sales in Thousands**

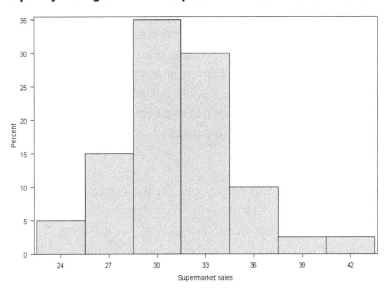

```
   Frequency Histogram for the Supermarket Cash
            Sales in Thousands

          The UNIVARIATE Procedure

         Histogram Bins for amount

                    Bin
            Minimum      Observed
             Point        Percent

              20          5.000
              25         32.500
              30         52.500
              35          7.500
              40          2.500
```

# 95% Confidence Interval

## Problem Setup

We use a random sample to represent the population as a whole because the random sample is an unbiased representation to the population. This property gives us a very powerful advantage to draw a conclusion about the population we want to study. If the size of the sample is sufficiently large, we are safe to infer about the population from the random sample.

Very often we want to know how likely a certain amount of sample average we can observe from the population. At here, "how likely" is a measure of probability. For instance, the average human body temperature is 38.6 °C; and if we got a sample averaging 39.9°C; we could ask, what probability or how likely is the randomly sampled average 39.9°C to occur in a usual population?

If the probability is very low; then we would be able to say it is very unlikely to obtain a sample of average body temperature 39.9°C from the usual human population and there must be something wrong with the sample.

A 95% confidence interval covers 95% of all possible sample averages. This means we have a 95% chance to get a sample mean fall in between the 95% confidence interval; and otherwise there is a 5% chance to have a sample average falling outside the 95% confidence interval. If we have a sample with average falls outside the 95% confidence interval, and the probability for this to happen is only 5%, which is very low; we are able to conclude that the sample is not likely a product of the usual population.

# SAS Code

```
options pageno=1 nodate nocenter;
data Temperature;
    input tem ;
    label tem='Temperature measure';
    datalines;
    17
    21
    18
    21
    19
    19
    19
    20
    19
    18
;
run;
title 'Confidence Interval for the Temperature
measure Data';
proc means  data=Temperature lclm uclm alpha=.05;
    var tem;
run;

ods select BasicIntervals;
proc univariate data=Temperature cibasic
(alpha=.05);
      var tem;
run;
```

# Interpreting the Result

```
Confidence Interval for the Temperature measure
                     Data

             The MEANS Procedure

   Analysis Variable : tem Temperature measure

         Lower 95%          Upper 95%
         CL for Mean        CL for Mean
      ƒƒƒƒƒƒƒƒƒƒƒƒƒƒƒƒƒƒƒƒƒƒƒƒƒƒƒƒƒƒƒƒƒƒ
        18.1795618          20.0204382
      ƒƒƒƒƒƒƒƒƒƒƒƒƒƒƒƒƒƒƒƒƒƒƒƒƒƒƒƒƒƒƒƒƒƒ

   Confidence Interval for the Temperature measure
                     Data

           The UNIVARIATE Procedure
      Variable:   tem   (Temperature measure)

    Basic Confidence Limits Assuming Normality
```

| Parameter | Estimate | 95% Confidence Limits | |
|---|---|---|---|
| Mean | 19.10000 | **18.17956** | **20.02044** |
| Std Deviation | 1.28668 | 0.88503 | 2.34898 |
| Variance | 1.65556 | 0.78327 | 5.51772 |

In the SAS code, we have alpha = 0.05, this means we want a 95% confidence interval. If we specify alpha = 0.1, we can have a 90% confidence interval.

The 95% confidence intervals of the average are highlighted. We can see the means procedure only gives confidence interval of the average. The univariate procedure gives confidence interval of the mean, standard deviation, and variance.

# T-Test

## Problem Setup

A statistical test is just a cutoff value. Suppose we performed repeated random samplings on a population to discover that 95% of the time the average ages of the repeated samples fell between 30 and 50 years. Another researcher performed yet another independent sample and came up with an average age of 29 years.

Use the cutoffs of 30 and 50 from our own study. We came to the conclusion that the chance of getting a random sample average of 29 years old is less than 5%. Therefore, we must question how a sample of 29 years as an average age could happen. The answer could vary, all the way from a biased sample, or that the researcher sampled a different population, or, it just happened by chance. Regardless, we have enough evidence to reject this researcher's study.

The SAS code on the next page has data of temperature measurements from a sample of liquid solutions. Someone claimed about H0=20, i.e. he believes the average temperature is 20. And from the actual sample we can test if the H0 is plausible or not. This is something different from above the average age case which we need repeated sampling to learn about the distribution of sampled average age. In this liquid solution case we do not need to do repeated sampling, it is because the SAS code performs a test call T-test.

# SAS Code

```
options nodate pageno=1 nocenter;
data temperature;
input tem @@;
datalines;
    17 19
    21 19
    18 20
    21 19
    19 18
;
run;
proc ttest data=temperature H0=20;
     var tem;
run;
```

# Interpreting the Result

```
Confidence Interval for the Temperature measure Data
1

The T-TEST Procedure

Variable:  tem

   N         Mean      Std Dev      Std Err     Minimum     Maximum
  10      19.1000       1.2867       0.4069     17.0000     21.0000

     Mean         95% CL Mean        Std Dev      95% CL Std Dev
  19.1000     18.1796  20.0204       1.2867       0.8850    2.3490

     DF     t Value      Pr > |t|
      9       -2.21        0.0543
```

H0: the average temperature is 20.

H1: the average temperature is not 20.

Level of significance: α=0.05

P-value = 0.0543 (It is the probability for H0 is true)

Because the p-value is greater than the level of significance, we failed to reject the null hypothesis. We do not have significant evidence to conclude that the average temperature is not 20.

# Paired T-Test

## Problem Setup

For instance, we have many identical mechanical devices and their times to finish a single revolution are measured twice. The first revolution time is measured under standard condition. The second revolution time is measured under a condition called xp, which is an increase of surrounding temperature. Because we suspect the temperature change can affect the precision of the device.

Each device is measured under standard condition and xp condition, and produces a pair of data. We are interested to know if the difference, calculated by subtracting standard condition timing by xp condition timing within each pair, is about 0. In other words, the precision is more likely not to be affected by surrounding temperature, if the conclusion is the difference is about 0.

The following SAS code specifies 2 columns of data, with each row of data being a pair. Inside the t-test procedure, we also specify the pair is between xp*standard. When SAS runs, it takes the difference within each pair; and uses the differences to do one sample t-test against 0. The alpha=0.05 is for constructing 95% confidence interval.

# SAS Code

```
options pageno=1 nodate nocenter;
data timing;
input xp standard;
datalines;
3.61   3.96
2.98   4.28
3.90   3.14
2.47   3.76
2.92   3.95
3.32   3.54
3.19   3.64
3.43   3.06
3.46   3.71
3.17   3.44

;
run;
proc ttest data=timing alpha=.05;
      paired xp*standard;
run;
```

# Interpreting the Result

```
                        The T-TEST Procedure

Difference:  xp - standard

   N        Mean       Std Dev      Std Err      Minimum      Maximum

  10       -0.4030      0.6678       0.2112      -1.3000       0.7600

     Mean          95% CL Mean        Std Dev       95% CL Std Dev

  -0.4030      -0.8808    0.0748       0.6678       0.4594    1.2192

     DF      t Value       Pr > |t|

      9       -1.91         0.0887
```

H0: the average of single revolution time has no difference with the 2 treatments, i.e. the average difference is 0.

H1: there is a difference of the single revolution time under the 2 treatments.

Level of significance: α=0.05

P-value = 0.0887

Because the p-value is greater than 0.05, we failed to reject the H0. We have no significance evidence to support the device can have different revolution time under the 2 treatments.

# QQ Plot

## Problem Setup

Very often with a given set of data, we want to know if it is plausible to say the data is normally distributed. QQ plot is the standard way to do that.

It plots rank based z-score, in other words normal quantile, against original data. If the plot appears to have a linear pattern or increases following a straight line; then it is plausible to say the data is normally distributed.

Personally I like to use the term of Rank based Z-score; because it tells us how to do the normal quantile calculation which means we use data ranks and transform them into z-scores.

## SAS Code

```
options nodate pageno=1 nocenter;
data temperature;
      input tem @@;
      datalines;
      17 19
      21 19
      18 20
      21 19
      19 18
      ;
run;

symbol color=black v=plus;
title 'Normal Quantile-Quantile Plot for Tem Data';
proc univariate data=temperature noprint;
      qqplot tem / normal(mu=est sigma=est
      color=black l=2 w=2)
      square cframe = white;
run;
```

# Interpreting the Result

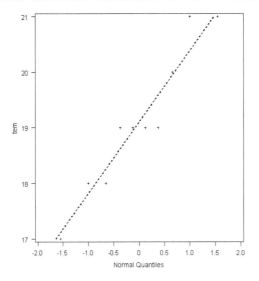

For the QQ plot we can see a dashed black line, it is the straight line we could observe if we plot perfectly normal distributed data. If the sample data is distributed relatively normal, we should see them closely scattered along the black line.

The blue colored plus signs are the sample quantiles. And we can see a straight line pattern from the blue plus signs because they follow the black line closely. We can conclude that it is plausible the sample data are normally distributed.

# T-Test with Two Treatment Groups

## Problem Setup

There is a type of material and we want to study the time it takes for this material to wear out. We have two different treatments for the material - treatment 1 and 2. We also want to determine if one method of treated material is lasting significantly longer than the other.

After one treatment we will not use the same materials to test the other treatment, because they have worn out. However, the paired t-test, as we discussed before, applied both treatments on the same test subject.

In this SAS code, data has 2 columns; with the first indicating the treatment groups of the material, and the second column providing data on the duration time of each material tested.

One significant twist is that data from the two treatment groups can have unequal variances. According to this test, the calculation is different for equal and unequal variance data. Therefore the t-test procedure also computes a statistic to tell us if it is plausible to say the data from two treatment groups have equal variance.

# SAS Code

```
goptions ls=72;
data time;
input method $ measure ;
cards;
1      3.42
1      2.43
1      2.73
1      2.96
1      3.12
1      4.15
1      2.61
1      2.57
1      2.54
1      2.62
2      3.16
2      3.41
2      3.38
2      3.63
2      3.70
2      3.87
2      4.20
2      3.99
2      3.24
2      3.44


;

proc ttest data=time;
      class method;
      var measure;
run;
```

# Interpreting the Result

```
The TTEST Procedure

Variable:  measure

Method            Mean      95% CL Mean            Std Dev     95% CL Std Dev

1                2.9150    2.5355   3.2945        0.5306      0.3649   0.9686
2                3.6020    3.3601   3.8439        0.3382      0.2326   0.6173

Method            Variances        DF    t Value    Pr > |t|

Pooled           Equal            18      -3.45      0.0028
Satterthwaite    Unequal      15.276      -3.45      0.0035

                 Equality of Variances

Method      Num DF    Den DF    F Value    Pr > F

Folded F        9         9       2.46     0.1958
```

First of all, the probability of equal variance is very high at 0.1958. We do not have enough evidence to say the variances of 2 treatment groups are significantly different. So we look at the testing treatment effect under equal variance conditions, i.e. pooled.

H0: the average duration time for the material is no different between the 2 treatment groups.

H1: There is a difference in the duration time of the material between the 2 treatment groups.

Level of significance $\alpha$=0.05, P-value = 0.0028

Because the p-value is less than the level of significance $\alpha$=0.05, we are able to reject the H0. There are significant evidences to support that the average duration time of the 2 treatment groups are different. Mean of first treatment is 2.915; mean of second treatment is 3.6020. It takes more time for the material of second treatment of to wear out.

# Wilcoxon Test

## Problem Setup

The previous t-test for comparing two treatment groups has a vital assumption: the data distributions from the two groups must be normal.

What if the data is not normally distributed and we confirmed the non-normality by looking at their QQ plot. The Wilcoxon test is specially designed to make comparisons for this type of situation.

The Wilcoxon test tells the difference by comparing data distributions. If a statistic of Wilcoxon test is significant, then the distributions of the data groups are significantly different.

# SAS Code

```
data time;
      input method measure ;
      datalines;
1     3.42
1     2.43
1     2.73
1     2.96
1     3.12
1     4.15
1     2.61
1     2.57
1     2.54
1     2.62
2     3.16
2     3.41
2     3.38
2     3.63
2     3.70
2     3.87
2     4.20
2     3.99
2     3.24
2     3.44
;
run;

proc npar1way data=time wilcoxon;
      class method;
      var measure;
      exact wilcoxon;
run;
```

# Interpreting the Result

```
        The NPAR1WAY Procedure
      Wilcoxon Two-Sample Test

Statistic (S)                 68.0000

Normal Approximation
Z                             -2.7591
One-Sided Pr <   Z             0.0029
Two-Sided Pr >  |Z|            0.0058

t Approximation
One-Sided Pr <   Z             0.0062
Two-Sided Pr >  |Z|            0.0125

Exact Test
One-Sided Pr <=   S            0.0019
Two-Sided Pr >=  |S - Mean|    0.0039

Z includes a continuity correction of 0.5.
```

H0: the distributions of the data from the 2 treatment groups are identical.

H1: the distributions of the data from the 2 treatment groups are not identical.

Level of significance: $\alpha=0.05$

P-value = 0.0039 (two-sided for testing equality)

Because the p-value is significant less than the level of significance, we are able to reject the H0. We have very strong evidence to support that the distributions of the data from the 2 treatment groups are not identical.

# One-Way ANOVA

## Problem Setup

ANOVA stands for "analysis of variance". We have discussed how to compare the differences between two treatment groups. What if we need to compare data from more than 2 treatment groups? The answer is one-way ANOVA.

For instance, we have 4 kinds of additives for special purpose oil. The additives are supposed to improve light transparency, and we want to discover if some of the additives are significantly better than the other.

We make 20 identical oil samples and divide them in to 4 groups. Each group we treat with one type of additive. We then measure and record the light transparency of each oil sample. The SAS code data has 2 columns, with the first indicating treatment group, and the second indicating the light transparency measurement.

# SAS Code

```
options nodate pageno=1 nocenter;
data oil;
input trt md @@;
datalines;
1      20.9
1      17.6
1      20.4
1      19.7
1      15.7
2      12.3
2      14.8
2      17.0
2      12.5
2      13.6
3      14.1
3      14.2
3      13.4
3      12.1
3      11.2
4      15.6
4      11.4
4      13.0
4      12.3
4      11.9

;
run;
proc means data=oil var;
     var md;
     by trt;
run;
proc anova data=oil;
     class trt;
     model md = trt;
     means trt / bon hovtest welch;
run;
```

# Interpreting the Result

```
                         The ANOVA Procedure

Dependent Variable: md
                                        Sum of
Source              DF        Squares   Mean Square   F Value   Pr > F
Model                3    120.4495000    40.1498333     12.47   0.0002
Error               16     51.4960000     3.2185000

Corrected Total     19    171.9455000

R-Square      Coeff Var      Root MSE      md Mean
0.700510       12.21667      1.794018     14.68500

Source              DF       Anova SS   Mean Square   F Value   Pr > F
trt                  3    120.4495000    40.1498333     12.47   0.0002
```

H0: no difference among the treatment population means

H1: some of the treatment population means differ from each other.

Level of significance: α = 0.05

P-value = 0.0002

Because the p-value is less than the level of significance, we are able to reject the H0. We have very strong evidence to conclude that some of the treatment groups show a mean measurement that significantly differs from the others.

# Two-Way ANOVA

## Problem Setup

We have seen that each test sample is assigned to only 1 treatment in the One-Way ANOVA example. However it is possible to give a test sample 1 or more treatments; and the treatments also can interact with each other to affect the test result in many different ways.

For instance, we want to test different formulations of an industrial chemical. The chemical is composed of 2 elements - first the ingredient and, second the base. The concentration of the ingredient we use can be low or high. There are also 3 types of difference bases available for us. By considering different combinations of ingredient and base, we have a total of 6 formulations for the industrial chemical.

We tested each formulation twice, allowing us to examine the interactive effect between ingredient and base. As listed in the SAS code, the first column indicates the ingredients used while the second indicates the bases used. Remember: we went through each formulation twice.

In this case each formulation is considered by 2 factors - the ingredient and the base. Hence, we call this type of analysis Two-Way ANOVA. We call the experimental design with 2 or more factors, the Factorial Design.

# SAS Code

```
options nodate pageno=1 nocenter;
data chemo;
input ingre base $ measure;
datalines;
1      1       107
1      1       112
1      2       113
1      2       116
1      3       132
1      3       188
2      1       126
2      1       143
2      2       148
2      2       176
2      3       236
2      3       222
;
run;
proc sort;
      by ingre base ;
run;
proc means data=chemo ;
      var measure;
      by ingre base ;
      output out = chemomeans mean = meanchemo;
run;

proc anova data=chemo; * with bonferroni interval;
      class ingre base;
      model measure = ingre base ingre*base;
      *also testing interaction;
      means ingre base / cldiff bon;
run;
```

# Interpreting the Result

The ANOVA Procedure

Dependent Variable: measure

| Source | DF | Sum of Squares | Mean Square | F Value | Pr > F |
|---|---|---|---|---|---|
| Model | 5 | 19221.41667 | 3844.28333 | 10.39 | 0.0064 |
| Error | 6 | 2219.50000 | 369.91667 | | |
| Corrected Total | 11 | 21440.91667 | | | |

| R-Square | Coeff Var | Root MSE | measure Mean |
|---|---|---|---|
| 0.896483 | 12.68821 | 19.23322 | 151.5833 |

| Source | DF | Anova SS | Mean Square | F Value | Pr > F |
|---|---|---|---|---|---|
| ingre | 1 | 6674.08333 | 6674.08333 | 18.04 | 0.005 |
| base | 2 | 11579.16667 | 5789.58333 | 15.65 | 0.0042 |
| ingre*base | 2 | 968.16667 | 484.08333 | 1.31 | 0.3376 |

There are 3 pairs of hypothesis we need to test: the main effect of ingredient, the main effect of base, and test interaction between ingredient and base.

H0: the ingredients have no effect on the function of the chemical.
H1: some ingredients made effect on the function of the chemical.
$\alpha=0.05$, p-value=0.0054, we are able to reject H0, we have very strong evidence to say there are ingredients that can have major effect on the function of the chemical.

H0: the bases have no effect on the function of the chemical.
H1: some bases made effect on the function of the chemical.
$\alpha=0.05$, p-value=0.0042, we are able to reject H0, and we have very strong evidence to say there are bases that can have major effect on the function of the chemical.

H0: the interactions between the ingredients and bases have no effect on the function of the chemical.
H1: some interactions between the ingredients and bases had effect on the function of the chemical.
$\alpha=0.05$, p-value=0.3376, we are not able to reject H0 and we have no evidence to say the interactions between the ingredients and bases can have any effect on the function of the chemical.

# Fisher Exact Test

## Problem Setup

Fisher exact test is a test for contingency tables, it tests frequencies among categories. Keep in mind, T-test and ANOVA compare means of random samples. Wilcoxon test compares distributions of random samples.

For instance, we want to test if the cooking time significantly affects the taste of some kinds of food. There are 2 cooking times; 4 and 6 minutes. Taste options are hard and soft. We made 11 different cookings of 4 minutes, and another 11 cookings of 6 minutes.

The 2x2 contingency table looks like this:

|        | Hard | Soft |
|--------|------|------|
| 4 mins | 3    | 8    |
| 6 mins | 9    | 2    |

## SAS Code

```
data food;
input  time taste $ count;
datalines;
4 hard 3
4 soft 8
6 hard 9
6 soft 2
;
run;
proc freq data=food;
weight count;
table time*taste;
exact fisher;
run;
```

# Interpreting the Result

```
           Table of time by taste
      time        taste

      Frequency,
      Percent  ,
      Row Pct  ,
      Col Pct  ,hard     ,soft      ,   Total
      fffffffff^fffffffff^fffffffff^
           4 ,        3 ,        8 ,      11
             ,    13.64 ,    36.36 ,   50.00
             ,    27.27 ,    72.73 ,
             ,    25.00 ,    80.00 ,
      fffffffff^fffffffff^fffffffff^
           6 ,        9 ,        2 ,      11
             ,    40.91 ,     9.09 ,   50.00
             ,    81.82 ,    18.18 ,
             ,    75.00 ,    20.00 ,
      fffffffff^fffffffff^fffffffff^
      Total           12        10        22
                   54.55     45.45    100.00

             Fisher's Exact Test
      fffffffffffffffffffffffffffffffff
      Cell (1,1) Frequency (F)          3
      Left-sided Pr <= F           0.0150
      Right-sided Pr >= F          0.9990

      Table Probability (P)        0.0140
      Two-sided Pr <= P            0.0300

             Sample Size = 22
```

H0: two treatments or the cooking times have equal effect on the taste of the food.

H1: two treatments or the cooking times have different effect on the taste of the food.

$\alpha=0.05$

P-value = 0.0300

Because the p-value is less than the level of significance, we are able to reject the H0. We can conclude that there is significant evidence to support the theory that the 2 cooking times have different effect on the taste of the food.

# Chi Square Test

## Problem Setup

When we have contingency with more than 2 columns and 2 rows, we use the Chi Square test. For instance the table below:

|  | Appreciable loss | Little change | increase |
|---|---|---|---|
| Control | 43 | 20 | 12 |
| Therapy | 27 | 88 | 21 |
| Activity | 20 | 35 | 30 |

The study tries to determine if incidences of weight change are the same by using 3 weight loss methods. The first is a control method, requiring participants to do nothing to their weight – it is life as usual. The second method is a therapeutic approach for managing weight. The third involves the use of activity to change people's weight.

It is very important to notice that the number of people for each weight loss method group is fixed by the study. We can pre-determine the number of people we want to assign to each group. So the row sums of the table are not random. Column sums, however, are random, meaning they are not fixed by the study.

In order for the Chi Square test to work properly, the cell count for each data cell of the contingency table cannot be less than 5.

# SAS Code

```
data weight;
input group $ level $ count;
datalines;
Control     Appreciable_loss                43
Control     Little_change                   20
Control     increase                        12
Therapy     Appreciable_loss                27
Therapy     Little_change                   88
Therapy     increase                        21
Activity    Appreciable_loss                20
Activity    Little_change                   35
Activity    increase                        30
;
run;
proc freq data=weight;
     weight count;
     table group*level / chisq;
run;
```

# Interpreting the Result

```
                          The FREQ Procedure
                     Table of group by level

group        level
Col Pct    ,Apprecia,Little_c,increase,  Total
fffffffff^fffffffff^fffffffff^fffffffff^
Activity ,     20 ,     35 ,     30 ,     85
         ,   6.76 ,  11.82 ,  10.14 ,  28.72
         ,  23.53 ,  41.18 ,  35.29 ,
         ,  22.22 ,  24.48 ,  47.62 ,
fffffffff^fffffffff^fffffffff^fffffffff^
Control  ,     43 ,     20 ,     12 ,     75
         ,  14.53 ,   6.76 ,   4.05 ,  25.34
         ,  57.33 ,  26.67 ,  16.00 ,
         ,  47.78 ,  13.99 ,  19.05 ,
fffffffff^fffffffff^fffffffff^fffffffff^
Therapy  ,     27 ,     88 ,     21 ,    136
         ,   9.12 ,  29.73 ,   7.09 ,  45.95
         ,  19.85 ,  64.71 ,  15.44 ,
         ,  30.00 ,  61.54 ,  33.33 ,
fffffffff^fffffffff^fffffffff^fffffffff^
Total          90      143      63      296
             30.41    48.31    21.28   100.00


Statistic                     DF      Value     Prob
ffffffffffffffffffffffffffffffffffffffffffffffffffffffff
Chi-Square                     4     50.9280   <.0001
Likelihood Ratio Chi-Square    4     47.7646   <.0001
Mantel-Haenszel Chi-Square     1      1.1892   0.2755
Phi Coefficient                       0.4148
Contingency Coefficient               0.3831
Cramer's V                            0.2933
```

H0: The proportion of the 3 weight loss levels are identically distributed for each treatment group.

H1: The proportion of the 3 weight loss levels are not identically distributed for each treatment group.

Level of significance: $\alpha = 0.05$

P-value < 0.0001

Because the p-value is significantly less than $\alpha$, we are able to reject the H0. We have very strong evidence to conclude that the 3 weight loss levels' distributions are not identically distributed among different treatment groups. There is strong association between weight loss level and the treatment for which the subjects were assigned.

# Linear Regression

## Problem Setup

Very often the change of one measurement can depend on the change of another measurement at a relatively constant rate. For example: the GPA of a student depends on how many hours the student spent in study; and on average as students spend longer hours on their studies, they achieve higher GPA. Another example: the price of a second hand car depends on its age; as the second hand cars ages, its average re-sells price drops.

The purpose of Linear Regression is to use one measurement to model the average value of another measurement. Regression means the tendency towards the average; for instance, height of a child tends to be the average height of the parents. Linear means the straight line model.

The SAS code gives us an example of second hand car prices. We can see the plot of SAS result, where price depends on the age in a decreasing straight line fashion. SAS also provides the slope and intercept of this decreasing straight line. We can now use the elementary Algebra knowledge; given a specific age of cars we can get a predicted average price of the cars at that age, using age × slope + intercept.

## SAS Code

```
data car;
input age price;
cards;
5      14995
9      7500
12     2900
5      15995
3      15000
2      20777
3      17931
6      11995
13     1200
7      9995
;

run;

symbol1 color=black i=none v=dot height=0.5;

proc gplot data=car;
     plot price*age =1 ;
run;
proc reg data=car;
     model price =age / p ;
     output out=car_reg p=predicted r=residual;
run;
```

# Interpreting the Result

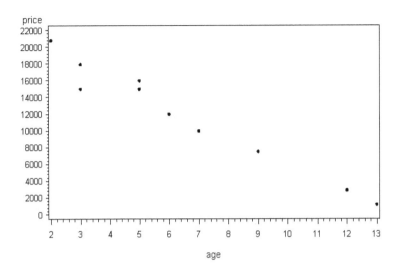

```
                    The SAS System
                    The REG Procedure

Root MSE              1325.04041    R-Square    0.9620
Dependent Mean           11829      Adj R-Sq    0.9572
Coeff Var             11.20182

                    Parameter Estimates

Variable    DF      Estimate        Error    t Value    Pr > |t|
Intercept   1         22641     867.66756      26.09     <.0001
age         1    -1663.34630     116.89005     -14.23     <.0001
```

Let "y" be the predicted average car price, "x" is the age of car. Also let "a" be the intercept of the line equation, and "b" be the slope.

$$y = a + bx$$

From the SAS output above, we can have a linear equation for the predicated average car price at a certain car age.

$$y = 22641 + (-1663.34630)x$$

We also need to know if the intercept and slope are significant.

H0: intercept of the linear regression model is 0.
H1: intercept of the linear regression model is not 0.
Level of significance: $\alpha = 0.05$
P-value < 0.0001
P-value is significantly less than $\alpha$. We are able to reject the H0. We have strong evidence to support that the intercept of the linear regression model is not 0.

H0: slope of the linear regression model is 0.
H1: slope of the linear regression model is not 0.
Level of significance: $\alpha = 0.05$
P-value < 0.0001
P-value is significantly less than $\alpha$. We are able to reject the H0. We have strong evidence to support that the slope of the linear regression model is not 0.

# Kaplan Meier Survival Curve

## Problem Setup

Consider the medical trial to study efficacy of treatments aimed at improving survival time for patients with a fatal disease. Suppose there are 2 different treatments we need to compare. On a certain day we start to recruit patients; it is impossible to have a sufficient number of patients on the same day, and because the disease occurs by chance, we can only wait for qualified patients.

Randomly assign qualified patients to the 2 treatment groups and measure how long the treatments have kept each patient alive. After a period of time the trial meets its exit criteria we stop the trial and calculate the duration of survival. Exit criteria mean the criteria for the entire experiment to stop.

Some of the patients recruited during the trial may still be alive when the trial finishes. We do a count of those patients, calling this final count "censoring". The term "got censored" means a patient survived the trial and made the final count.

Those not able to survive the trial were recruited on separate days and would have their different survival times during the trial. The survival times of those who were censored came between recruitment and the end of the trial.

In the SAS code, time variable gives the survival time of each patient. If a patient is censored, we assign "1" to the censor variable, otherwise "0". The group variable indicates which treatment group the patients are assigned to.

## SAS Code

```
options ls=72;
data one;
input time censor group;
cards;
6       0       1
7       0       1
6       0       1
8       1       1
8       0       1
16      0       1
20      1       1
21      0       1
25      0       1
25      1       1
25      0       1
26      0       1
26      0       1
27      0       1
3       1       2
6       0       2
6       0       2
6       0       2
5       1       2
8       0       2
9       0       2
6       0       2
8       0       2
8       0       2
16      0       2
21      0       2
19      1       2
19      0       2
21      0       2
22      0       2
28      0       2
;
proc lifetest plots=(s);
        time time*censor(1);
        strata group;
        title 'Survival Curve';
        label group='treatmnet group'
```

```
        time='death time';
run;
```

# Interpreting the Result

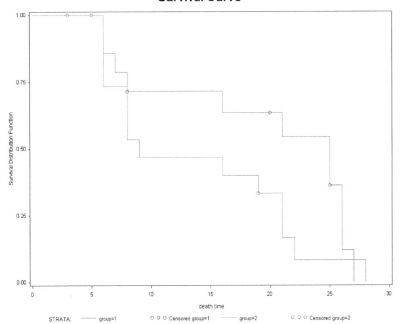

**Survival Curve**

```
Test of Equality over Strata

                                    Pr >
   Test       Chi-Square    DF    Chi-Square

   Log-Rank     1.9219       1      0.1656
   Wilcoxon     2.3458       1      0.1256
   -2Log(LR)    0.9684       1      0.3251
```

The statistical test to compare two Kaplan Meier Survival Curves is called Log-Rank test. The SAS output gives us p-value of this test.

HO: survival time of the two survival curves is identical.
H1: survival time of the two survival curves is not identical.

Level of significance: $\alpha = 0.05$
P-value = 0.1656

Because the p-value of the Log-Rank test is higher than $\alpha$ which we set. We failed to reject the HO. We have no evidence to conclude that the survival time of the two survival curves are not identical. That being so, it is plausible to say the two survival curves are equal.

# Cochran-Mantel-Haenszel (CMH) Test

## Problem Setup

CMH test is for retrospective studies, which means studies look back on things already happened. We do not need precise measurements to perform this test. We do need, however, a set of categories which have some sort of increasing or decreasing order.

For instance, there is an eye exam for students. Some passed and some did not. Every student rated his computer daily use as well. From the data listed below we want to know if computer use has any significant effect on students' eye exam performance.

|  | Not much 0 | Less than average 0.5 | Usual 1.5 | More than usual 3 | A lot 6 |
|---|---|---|---|---|---|
| Passed | 575 | 488 | 30 | 10 | 7 |
| Didn't pass | 6 | 8 | 7 | 6 | 7 |

We do not want to assign students to specific computer operating hours randomly, because that is not something ethical to do. If we divide students into groups, and for each group we ask the students to use computer with predefined amount of time each day; we might compromise some students' ability to see. So we need the CMH test.

Note: to analyze the data we assign numerical value to computer usage categories just to tell SAS there is an increasing order. Categories: Not much is 0, Less than average is 0.5, Usual is 1.5, More than usual is 3, A lot is 6.

## SAS Code

```
data computer_hour;
input pass hour count;
cards;
1      0      575
1      0.5    488
1      1.5    30
1      3      10
1      6      7
2      0      6
2      0.5    8
2      1.5    7
2      3      6
2      6      7
;
proc freq;
      weight count;
      tables pass*hour / chisq cmh1;
run;
```

# Interpreting the Result

The FREQ Procedure

Table of pass by hour

```
pass       hour

Frequency,
Percent  ,
Row Pct  ,
Col Pct  ,        0,     0.5,     1.5,       3,       6,   Total
fffffffff^fffffffff^fffffffff^fffffffff^fffffffff^fffffffff^
        1 ,     575 ,     488 ,      30 ,      10 ,       7 ,    1110
          ,   50.26 ,   42.66 ,    2.62 ,    0.87 ,    0.61 ,   97.03
          ,   51.80 ,   43.96 ,    2.70 ,    0.90 ,    0.63 ,
          ,   98.97 ,   98.39 ,   81.08 ,   62.50 ,   50.00 ,
fffffffff^fffffffff^fffffffff^fffffffff^fffffffff^fffffffff^
        2 ,       6 ,       8 ,       7 ,       6 ,       7 ,      34
          ,    0.52 ,    0.70 ,    0.61 ,    0.52 ,    0.61 ,    2.97
          ,   17.65 ,   23.53 ,   20.59 ,   17.65 ,   20.59 ,
          ,    1.03 ,    1.61 ,   18.92 ,   37.50 ,   50.00 ,
fffffffff^fffffffff^fffffffff^fffffffff^fffffffff^fffffffff^
Total           581      496       37       16       14     1144
              50.79    43.36     3.23     1.40     1.22   100.00
```

Statistics for Table of pass by hour

```
Statistic                        DF       Value      Prob
ffffffffffffffffffffffffffffffffffffffffffffffffffffffffffff
Chi-Square                        4    216.9030    <.0001
Likelihood Ratio Chi-Square       4     80.8727    <.0001
Mantel-Haenszel Chi-Square        1    190.4196    <.0001
Phi Coefficient                           0.4354
Contingency Coefficient                   0.3992
Cramer's V                                0.4354
```

WARNING: 30% of the cells have expected counts less
         than 5. Chi-Square may not be a valid test.

Sample Size = 1144

H0: the correlation coefficient between the daily computer operating hours and the eye exam failed incidents is 0.
H1: the correlation coefficient between the daily computer operating hours and the eye exam failed incidents is not 0.

Level of significance: $\alpha = 0.05$
P-value < 0.0001

Because the p-value is significantly less than $\alpha$, we are able to reject H0. We have strong evidence to conclude that there is a correlation between daily computer operating hours and incidents of failed eye exam.

# Sample Size Calculation for Continuous Data

## Problem Setup

Suppose you are a statistician and there is a medical researcher asks for your help in calculating a sample size that is needed to compare a drug effect from two treatment groups. To one group we give the actual drug while the other receives a placebo. The drug is expected to produce a type of fatty acid level in the bloodstream.

Certainly you need more information to make the calculation. You asked: "How big is the difference you consider to be significant of fatty acid level between the two groups? How about 0.1 units"? You also offered him your suggestion. Then the medical researcher informed you: "0.1 units' difference is too small to be meaningful because the amount of this chemical in the bloodstream changes rapidly under various conditions such as food intake, exercise. We are looking at least 5 units' difference."

You now have a very important piece of information referred to as "minimal relevant difference" (MRD). Statistically, as the sample we use gets large, the ability to detect smaller differences also increases. Very often we do not need a very large sample size. Since the extremely small difference, in this case, is not really meaningful, we only need the sample size large enough to show the minimal relevant difference.

Power is another piece of information we need to specify. Think about it. If the minimal 5 units' difference is true between the 2 groups; there is always a chance that we do not detect any difference, simply because the study is randomized. The power is the probability we can detect the MRD. That, of course, hinges on the condition the MRD is true.

# SAS Code

```
*sample size of various powers, two-sided test;
data ss;
      do power = 0.80, 0.85, 0.90;
      alpha = 0.05;
      zalpha = probit(1-alpha/2);
      zbeta=probit(power);

      mu1 = 100; *mean fatty acid level in Group 1;
      diff = 5; *minimal relevant difference;

      *mean fatty acid level in Group 2;
      mu2 = mu1 - diff;

      *standard deviation of fatty acid level;
      sd = 10;

      numerator = zalpha*sd+zbeta*sd;

      *n is sample size of a single group;
      n = 2*(numerator**2)/(diff**2);
      output;
end;
run;

proc print;
      var power diff n;
run;
```

# Interpreting the Result

| Obs | power | diff | n |
|-----|-------|------|---------|
| 1 | 0.80 | 5 | 62.7910 |
| 2 | 0.85 | 5 | 71.8272 |
| 3 | 0.90 | 5 | 84.0594 |

We can always detect smaller treatment differences by using a larger sample size. If we want to detect only a certain amount of treatment difference, we do not need to make the sample size arbitrarily large. For this case, in order to make the treatment difference meaningful, we specify the minimal treatment difference or MRD as 5.

Power is the probability for us to be able to detect the MRD, if there is actually that specified amount of MRD. Because the sample is random and if the MRD is true, there is still a chance we will not be able to see the MRD from the sample. However, larger sample size can give us better probability to confirm the MRD, if it is true.

Here we specify MRD as 5, and we make calculations with 3 different powers. So that, each sample size "n" corresponds to a specified power. We can observe that as the power we desire increases, the sample size we need also increases.

And the sample size "n" is the number of test subjects of 1 single treatment group. As we planned to compare 2 treatment groups, we need a total sample size of 2n.

# Power Calculation for Continuous Data

## Problem Setup

Let's say the medical researcher came to you with another question. He wants to know the power of his drug trial, i.e. if the drug is able to make the minimum amount of effect to be significant. He wants to know the probability of the trial to confirm the existence of the effect.

From the previous sample size calculation, you know the sample size we need is just a function of power and minimal relevant difference. Then we can do the inverse function of sample size calculation. We derive the power is a function of sample size and minimal relevant difference.

## SAS Code

```
*calculate power for various sample sizes;
data power;

        *n is sample size of a single group;
        do n = 60, 70, 80;
                alpha = 0.05;
                zalpha = probit(1-alpha/2);

                *mean blood fatty acid in Group 1;
                mu1 = 100;

                diff = 5; *MRD;

                *mean blood fatty acid in Group 2;
                mu2 = mu1 - diff;

                *standard deviation of fatty acid;
                sd = 10;

                *std error under alternative;
                sealt = sqrt(2*sd**2/n);

                zbeta = diff/ sealt- zalpha ;
                power = probnorm(zbeta);
                output;
        end;
run;

proc print;
        var n diff power;
run;
```

# Interpreting the Result

```
Obs     n     diff     power

 1     60      5      0.78191
 2     70      5      0.84088
 3     80      5      0.88538
```

We still specify the MRD is 5. We can calculate the probability to confirm the MRD, if the MRD is true by given different sample sizes. As sample size increases, the probability or power to confirm the MRD also increases.

# Minimal Relevant Difference for Continuous Data

## Problem Setup

Let's say the medical research has picked his sample and he also has a preferable power in mind. You can tell the level of minimal relevant difference his trial is able to produce. Because by transforming the function of sample size calculation, we get minimal relevant difference as a function of sample size and power.

If the minimal relevant difference from your calculation is smaller than the medical researcher's need. You can tell the medical researcher to reduce the number of test subjects and save some budget.

And if the minimal relevant difference from your calculation is larger than the medical researcher's need. You can tell the medical researcher to be prepared for recruiting more test subjects because the sample size he provided was not enough.

# SAS Code

```
*detectable differences for different sample sizes;
data diff;
     do n = 40 to 80 by 20;
            alpha = 0.05;
            zalpha = probit(1-alpha/2);
            power = 0.80;
            zbeta = probit(power);

            *mean blood fatty acid in Group 1;
            mu1 = 100;

            *minimal relevant difference;
            diff = 5;

            *mean blood fatty acid in Group 2;
            mu2 = mu1 - diff;

            *standard deviation of fatty acid;
            sd = 10;

            *std error under alternative;
            sealt = sqrt(2*sd**2/n);

            diff = (zalpha+ zbeta)*sealt ;
            output;
     end;
run;

proc print;
     var n power diff;
run;
```

# Interpreting the Result

| Obs | n | power | diff |
|-----|-----|-------|---------|
| 1 | 40 | 0.8 | 6.26453 |
| 2 | 60 | 0.8 | 5.11497 |
| 3 | 80 | 0.8 | 4.42970 |

Let's explain them in sentences.

By using sample size 40 for each treatment group, the MRD we can detect with 80% chance is 6.26453.

By using sample size 60 for each treatment group, the MRD we can detect with 80% chance is 5.11497.

By using sample size 80 for each treatment group, the MRD we can detect with 80% chance is 4.42970.

# Sample Size Calculation for Proportion

## Problem Setup

Let's think about medical trials of binary outcomes. Binary outcomes means there are only 2 types of outcomes of the trial. For instance, after a treatment the patients can be cured or not cured which means the treatment can either a success or failure. So the binary outcomes for a medical clinical trial could be success or failure, survival or death, and so on.

Suppose we have a new cancer treatment that needs to be compared with the conventional cancer treatment. From previous studies we know there is 41% patients' cancer patients got worse after conventional treatment. If the new experimental treatment is better, we expect to see fewer cases of worsening cancer after the new treatment.

The minimal relevant difference is 27% for what we are looking at. In other words, the percentage for cases of worsening cancer after new treatment should be less than 14% in order for the new treatment to be medically significant.

The SAS code calculates sample size with 3 different powers for only 1 treatment group. So that the total number of patients we need to recruit for this trial is 2 times the sample size for 1 treatment group.

## SAS Code

```
*sample size for various power;
data ss;
     do power = 0.80, 0.85, 0.90;

             alpha = 0.05;
             zalpha = probit(1-alpha/2);
             zbeta=probit(power);

*probability of got worse on conventional
treatment;
             pi1 = 0.41;

*minimal relevant difference we wish to detect;
             diff = 0.27;

*probability of got worse on exp. treatment;
             pi2 = pi1 - diff;

*probability got worse under no difference;
             pibar = (pi1+pi2)/2;

numerator = zalpha*sqrt(2*pibar*(1-
pibar))+zbeta*sqrt( pi1*(1-pi1)+ pi2*(1-pi2));

             * sample size assumes n=n1=n2;
             n = (numerator**2)/(diff**2);
             output;
     end;
run;

proc print;
     var power diff n;
run;
```

# Interpreting the Result

| Obs | power | diff | n |
|-----|-------|------|---------|
| 1 | 0.80 | 0.27 | 41.7333 |
| 2 | 0.85 | 0.27 | 47.5331 |
| 3 | 0.90 | 0.27 | 55.3666 |

The progression of cancer needs to decrease by at least 27% before we can say the new treatment made a difference. As the probability to detect the difference grows, we need a larger sample size.

# Power Calculation for Proportion

## Problem Setup

Give the sample size and minimal relevant difference of the binary outcome clinical trial we can calculate the power for it.

The SAS code calculates 3 powers for 3 single treatment group sample sizes. With conventional treatment, we still have a 41% incidence of cancer patients whose condition is worsening. The new treatment, however, is at least 27% better than the conventional treatment.

# SAS Code

```
*calculate power for various sample sizes;

data power;
      do n = 60, 70, 80;
            alpha = 0.05;
            zalpha = probit(1-alpha/2);
            pi1 = 0.41;
            diff = 0.27;
            pi2 = pi1 - diff;
            pibar = (pi1+pi2)/2;

            *std error under null;
            senull = sqrt(2*pibar*(1-pibar)/n);

            *std error under alt;
sealt = sqrt(( pi1*(1-pi1)+ pi2*(1-pi2))/n);

            zbeta = (diff - zalpha*senull)/sealt;
            power = probnorm(zbeta);
            output;
      end;
run;
```

# Interpreting the Result

| Obs | n | diff | power |
|-----|-----|------|---------|
| 1 | 60 | 0.27 | 0.92196 |
| 2 | 70 | 0.27 | 0.95513 |
| 3 | 80 | 0.27 | 0.97476 |

If there exists the 27% difference, the probability to confirm the difference gets larger when the sample size gets larger.

# Minimal Relevant Difference for Proportion

## Problem Setup

For a binary outcome clinical trial if we already have the sample size for each treatment group, then by given specific levels of power we can determine the least amount of treatment difference or MRD for each power.

The larger sample size allows us to have more power and to be able detect smaller treatment difference. If we want more power with a fixed number of sample sizes; we will compromise the ability to detect smaller treatment difference, so the MRD gets larger. Large MRD means we can see only the relatively large or obvious treatment difference through our statistical test. If the true treatment difference is relatively small we might not be able to see it.

## SAS Code

```
*detectable differences for different sample sizes;
data diff;
      do n = 40 to 100 by 20;
            alpha = 0.05;
            zalpha = probit(1-alpha/2);
            power = 0.80;
            zbeta = probit(power);
            pi1 = 0.41;
            diff = 0.27;
            pi2 = pi1 - diff;
            pibar = (pi1+pi2)/2;

            *std error under null;
            senull = sqrt(2*pibar*(1-pibar)/n);

            *std error under alt;
sealt = sqrt(( pi1*(1-pi1)+ pi2*(1-pi2))/n);

            diff = zalpha*senull + zbeta*sealt;
            output;
      end;
run;

proc print;
      var n power diff;
run;
```

# Interpreting the Result

| Obs | n | power | diff |
|-----|-----|-------|---------|
| 1 | 40 | 0.8 | 0.27579 |
| 2 | 60 | 0.8 | 0.22518 |
| 3 | 80 | 0.8 | 0.19501 |
| 4 | 100 | 0.8 | 0.17442 |

If we require a fixed probability to detect the MRD under the condition the MRD is true, as sample size gets larger, the MRD we are able to detect gets smaller.

# Complete Randomization Allocation

## Problem Setup

Think about a clinical trial on testing the efficacy of a drug. We divide test subjects into 2 groups. We give the first group the actual drug as treatment while the second group is given a placebo. After a period of time, we compare treatment results from the 2 groups and try to conclude whether the drug is more effective than the placebo. In order to make this comparison valid, we must randomly assign subjects into the 2 groups.

Complete randomization allocation assigns subjects into 2 treatment groups by equal probability. For the sake of fairness, we toss a coin for each subject. If heads comes up, the subject goes to group A. If it is tails, the subject goes to group B. No doubt this allocation is unbiased.

Complete randomization allocation produces imbalanced groups sizes. The group sizes are not likely to be equal using this type of randomized allocation. Any difference in group size we call "imbalance". If the imbalance is large, one group size becomes really small and could not be representative enough for its treatment effect. We prefer to avoid imbalance on randomized allocation.

# SAS Code

```
data complete ;
      na=0;

      * 50 random allocations, for each of them;
      do i=1 to 50;

      * generate a random number between 1 - 0;
            x=ranuni(78968);

            * if the random number is > 0.5;
            if x>.5 then do;

                  t='A';       * allocate to A;
                  na+1;
            end;

            * otherwise allocate to B;
            else t='B';
            output;
      end;
run;

proc print;
      var i t na;
run;
```

# Interpreting the Result

| Obs | i | t | na |
|---|---|---|---|
| 1 | 1 | B | 0 |
| 2 | 2 | A | 1 |
| 3 | 3 | B | 1 |
| 4 | 4 | B | 1 |
| 5 | 5 | B | 1 |
| 6 | 6 | A | 2 |
| 7 | 7 | A | 3 |
| 8 | 8 | A | 4 |
| 9 | 9 | B | 4 |
| 10 | 10 | A | 5 |
| 11 | 11 | A | 6 |
| 12 | 12 | B | 6 |
| 13 | 13 | A | 7 |
| 14 | 14 | B | 7 |
| 15 | 15 | B | 7 |
| 16 | 16 | A | 8 |
| 17 | 17 | A | 9 |
| 18 | 18 | B | 9 |
| 19 | 19 | B | 9 |
| 20 | 20 | A | 10 |
| 21 | 21 | A | 11 |
| 22 | 22 | B | 11 |
| 23 | 23 | A | 12 |
| 24 | 24 | A | 13 |
| 25 | 25 | B | 13 |
| 26 | 26 | A | 14 |
| 27 | 27 | B | 14 |
| 28 | 28 | A | 15 |
| 29 | 29 | B | 15 |
| 30 | 30 | A | 16 |
| 31 | 31 | B | 16 |
| 32 | 32 | B | 16 |
| 33 | 33 | B | 16 |
| 34 | 34 | A | 17 |
| 35 | 35 | B | 17 |
| 36 | 36 | B | 17 |
| 37 | 37 | A | 18 |
| 38 | 38 | A | 19 |
| 39 | 39 | A | 20 |
| 40 | 40 | A | 21 |
| 41 | 41 | B | 21 |
| 42 | 42 | B | 21 |
| 43 | 43 | A | 22 |
| 44 | 44 | A | 23 |
| 45 | 45 | B | 23 |
| 46 | 46 | A | 24 |
| 47 | 47 | A | 25 |
| 48 | 48 | B | 25 |
| 49 | 49 | A | 26 |
| 50 | 50 | B | 26 |

Each test subject has 0.5 chances to be assigned to either of the two treatment groups under Complete Randomization Allocation. The number of test subjects in each treatment group is not likely to be equal. Complete Randomization Allocation is very likely to produce imbalanced test subject allocations.

From the outputted result, the column t indicates test subjects' treatment assignments. Column I is the test subjects' ID. Column na is the number of subjects have been allocated to treatment A. We can see after 50 test subjects assigning onto the 2 treatment groups under Complete Randomization Allocation, we have 26 subjects allocated to treatment A.

# Efron's Biased Coin Design

## Problem Setup

Efron's biased coin design is a randomized allocation scheme which tries to avoid imbalanced group sizes. Suppose we have 2 groups - A and B – to which we need to assign test subjects. The idea is to give more assignment probability to the smaller group.

We start with assignment by equal probability into the 2 group just like we did in complete randomization allocation. When number of subjects assigned to the 2 groups is equal, we continue assignment with equal probability. If the number of subjects assigned to the 2 groups is not equal, we continue assignment by giving higher probability to the group with fewer assigned subjects. Efron suggested the higher probability to be 2/3.

For instance, there are 50 subjects who need to be allocated into either group A or B. After 20 random assignments we counted 10 subjects in A and 10 subjects in B. Then the 21st random assignment has equal probability to go into A or B. If the 21st subject is assigned to B, now group B has 1 subject more than group A. Then the 22nd subject has 1/3 probability to go into B, and 2/3 probability to go into A.

# SAS Code

```
data efron;

nb=0;
na=0;

p=2/3; * Efron's suggestion of p;

do j=1 to 50; * There are 50 allocations;

      * When na=na, selection probability is 1/2;
      if nb-na=0 then phi=.5;

      * When there're too many B's, selection
probability is 2/3;
      else if nb-na>0 then phi=p;

      * When there're too many A's, selection
probability is 1/3;
      else phi=1-p;

      x=ranuni(3267);

      * If x< selection probability, allocate to A;
      if x<phi then do;
            t='A';
            na+1;
      end;

      * Otherwise allocate to B;
      else do;
            t='B' ;
            nb+1;
      end;
      output;
end;
run;
proc print;
      var j t na;
run;
```

# Interpreting the Result

| Obs | j | t | na |
|---|---|---|---|
| 1 | 1 | A | 1 |
| 2 | 2 | A | 2 |
| 3 | 3 | A | 3 |
| 4 | 4 | B | 3 |
| 5 | 5 | B | 3 |
| 6 | 6 | B | 3 |
| 7 | 7 | B | 3 |
| 8 | 8 | A | 4 |
| 9 | 9 | B | 4 |
| 10 | 10 | A | 5 |
| 11 | 11 | B | 5 |
| 12 | 12 | A | 6 |
| 13 | 13 | A | 7 |
| 14 | 14 | B | 7 |
| 15 | 15 | B | 7 |
| 16 | 16 | B | 7 |
| 17 | 17 | A | 8 |
| 18 | 18 | A | 9 |
| 19 | 19 | A | 10 |
| 20 | 20 | A | 11 |
| 21 | 21 | B | 11 |
| 22 | 22 | B | 11 |
| 23 | 23 | B | 11 |
| 24 | 24 | A | 12 |
| 25 | 25 | A | 13 |
| 26 | 26 | B | 13 |
| 27 | 27 | B | 13 |
| 28 | 28 | B | 13 |
| 29 | 29 | A | 14 |
| 30 | 30 | B | 14 |
| 31 | 31 | A | 15 |
| 32 | 32 | A | 16 |
| 33 | 33 | A | 17 |
| 34 | 34 | B | 17 |
| 35 | 35 | A | 18 |
| 36 | 36 | A | 19 |
| 37 | 37 | B | 19 |
| 38 | 38 | B | 19 |
| 39 | 39 | A | 20 |
| 40 | 40 | A | 21 |
| 41 | 41 | A | 22 |
| 42 | 42 | B | 22 |
| 43 | 43 | A | 23 |
| 44 | 44 | B | 23 |
| 45 | 45 | B | 23 |
| 46 | 46 | A | 24 |
| 47 | 47 | B | 24 |
| 48 | 48 | B | 24 |
| 49 | 49 | A | 25 |
| 50 | 50 | A | 26 |

Efron's way for us to allocate test subjects is that if we have an equal number of subjects on each treatment, then the next subject has equal probability to go into either treatment A or B; or if treatment A has more subjects already, the probability for the next subject go into group B is 2/3; or if treatment B has more subjects already, the probability for next subject go into group A is 2/3.

Efron suggested we give more allocation probability to the treatment group with fewer test subjects. Efron's suggestion still does not guarantee that we always have an equal number of test subjects for the treatment groups. From the outputted result we can see that, column na is for number of subjects allocated to treatment A, after 50 subjects' allocations we have 26 subjects in treatment A and 24 subjects in treatment B.

# Truncated Binomial Allocation

## Problem Setup

Truncated binomial allocation guarantees equal group size, meaning each group can take on half the total test subjects. We start with complete randomized allocation where subjects are assigned by equal probability. Once a group reaches half of the total subjects, we allocate the rest of the subjects to the other group.

For instance, there are 50 test subjects to be allocated to groups A and B. After 40 randomized assignments, group A has 25 subjects, and group B has only 15. We will allocate the remaining 10 subjects to group B.

## SAS Code

```
data truncated ;
          na=0;
          nb=0;
          M=50;          * There are 50 allocations;

     do i=1 to M; * For each allocation we do;

* number of A's reaches 25, only allocate to B;
          if na=M/2 then do;
                  t='B';
                  nb+1;
          end;

          *B's reaches 25 only allocate to A;
          if nb=M/2 and na<M/2 then do;

                  t='A';
                  na+1;
          end;

          * For na and nb not yet up to 25,;
          if max(na,nb)<M/2 then do;

                  x=ranuni(7309279);
                  if x>.5 then do;
                          t='A';
                          na+1;
                  end;
                  else do;
                          t='B';
                          nb+1;
                  end;
          end;

          output;
     end;
run;

proc print;
     var i t na nb;
run;
```

# Interpreting the Result

| Obs | i | t | na | nb |
|-----|-----|-----|-----|-----|
| 1 | 1 | B | 0 | 1 |
| 2 | 2 | B | 0 | 2 |
| 3 | 3 | B | 0 | 3 |
| 4 | 4 | B | 0 | 4 |
| 5 | 5 | B | 0 | 5 |
| 6 | 6 | A | 1 | 5 |
| 7 | 7 | B | 1 | 6 |
| 8 | 8 | B | 1 | 7 |
| 9 | 9 | B | 1 | 8 |
| 10 | 10 | B | 1 | 9 |
| 11 | 11 | B | 1 | 10 |
| 12 | 12 | A | 2 | 10 |
| 13 | 13 | B | 2 | 11 |
| 14 | 14 | A | 3 | 11 |
| 15 | 15 | B | 3 | 12 |
| 16 | 16 | A | 4 | 12 |
| 17 | 17 | A | 5 | 12 |
| 18 | 18 | B | 5 | 13 |
| 19 | 19 | A | 6 | 13 |
| 20 | 20 | A | 7 | 13 |
| 21 | 21 | A | 8 | 13 |
| 22 | 22 | B | 8 | 14 |
| 23 | 23 | A | 9 | 14 |
| 24 | 24 | B | 9 | 15 |
| 25 | 25 | A | 10 | 15 |
| 26 | 26 | A | 11 | 15 |
| 27 | 27 | A | 12 | 15 |
| 28 | 28 | B | 12 | 16 |
| 29 | 29 | A | 13 | 16 |
| 30 | 30 | A | 14 | 16 |
| 31 | 31 | A | 15 | 16 |
| 32 | 32 | A | 16 | 16 |
| 33 | 33 | A | 17 | 16 |
| 34 | 34 | A | 18 | 16 |
| 35 | 35 | A | 19 | 16 |
| 36 | 36 | A | 20 | 16 |
| 37 | 37 | B | 20 | 17 |
| 38 | 38 | A | 21 | 17 |
| 39 | 39 | B | 21 | 18 |
| 40 | 40 | A | 22 | 18 |
| 41 | 41 | A | 23 | 18 |
| 42 | 42 | A | 24 | 18 |
| 43 | 43 | B | 24 | 19 |
| 44 | 44 | B | 24 | 20 |
| 45 | 45 | A | 25 | 20 |
| 46 | 46 | B | 25 | 21 |
| 47 | 47 | B | 25 | 22 |
| 48 | 48 | B | 25 | 23 |
| 49 | 49 | B | 25 | 24 |
| 50 | 50 | B | 25 | 25 |

Truncated Binomial Allocation tells us that assign subjects to the 2 treatment groups by equal probability, if one of the treatment groups has reached half of the total number of subjects, and then allocate all the remaining subjects to the other treatment group.

Truncated Binomial Allocation always gives us an equal number of treatment allocations. From the outputted result we can see that once test subject 45 was assigned to treatment group A, that group at 25 subjects had enough. The remaining 5 subjects were allocated to B completely.

# Random Allocation Rule

## Problem Setup

Random allocation rule is a so-called urn design. Imagine we have an urn of balls, of which half are black, and half are white. Draw out 1 ball at a time without replacement. After we empty the urn, we obtain a randomized permutation of black and white balls. Using this sequence as a treatment group allocation is unbiased and balanced as well.

The SAS code on the following page simulates an urn. To randomize 50 test subjects, we first put 25 As in the urn, followed by 25 Bs. We associate each A and B with a randomly generated decimal number. We then sort by the decimal numbers. Since the decimal numbers are random, we randomize the permutation of As and Bs by sorting.

# SAS Code

```
data RAR;

      * There 50 random allocations;
      M=50;

      * There are 25 A's;
      do i =1 to M/2;
           t='A';
      * Each A gets a uniform(0,1) random number;
           order=ranuni(3267);
           output;
      end;

      * There are 25 B's;
      do i =1 to M/2;
           t='B';
      * Each B gets a uniform(0,1) random number;
           order=ranuni(267);
           output;
      end;
run;

* Sort A's and B's by the random number;
proc sort data=RAR;
      by order;
run;


proc print;
      var t order;
run;
```

# Interpreting the Result

| Obs | t | order |
|---|---|---|
| 1 | A | 0.00006 |
| 2 | A | 0.00010 |
| 3 | B | 0.01062 |
| 4 | B | 0.02704 |
| 5 | A | 0.03686 |
| 6 | A | 0.03921 |
| 7 | B | 0.04078 |
| 8 | B | 0.07373 |
| 9 | A | 0.09081 |
| 10 | B | 0.14464 |
| 11 | B | 0.14541 |
| 12 | B | 0.21559 |
| 13 | A | 0.22601 |
| 14 | A | 0.27272 |
| 15 | B | 0.27481 |
| 16 | B | 0.27931 |
| 17 | A | 0.27983 |
| 18 | B | 0.29385 |
| 19 | A | 0.36607 |
| 20 | A | 0.40366 |
| 21 | B | 0.42032 |
| 22 | B | 0.42612 |
| 23 | B | 0.43520 |
| 24 | A | 0.45537 |
| 25 | A | 0.46881 |
| 26 | A | 0.47292 |
| 27 | B | 0.47464 |
| 28 | B | 0.47552 |
| 29 | A | 0.48499 |
| 30 | A | 0.49201 |
| 31 | A | 0.49955 |
| 32 | B | 0.53563 |
| 33 | B | 0.55722 |
| 34 | B | 0.57245 |
| 35 | A | 0.59694 |
| 36 | A | 0.62271 |
| 37 | A | 0.62910 |
| 38 | A | 0.67708 |
| 39 | B | 0.70268 |
| 40 | A | 0.71377 |
| 41 | B | 0.72137 |
| 42 | A | 0.79629 |
| 43 | B | 0.83820 |
| 44 | B | 0.86566 |
| 45 | A | 0.87562 |
| 46 | B | 0.89966 |
| 47 | A | 0.91296 |
| 48 | A | 0.93268 |
| 49 | B | 0.96685 |
| 50 | B | 0.99029 |

We can guarantee equal number of treatment group allocation by putting equal numbers of A balls and B balls into the urn. We remove 1 ball each time without replacing it. Each ball has a random number called order. We sort the balls by the randomized order.

# Permuted Block Design

## Problem Setup

There is a type of sampling design called stratification. We divide the population into a few strata and randomly sample each stratum. This kind of random sampling is also unbiased and guarantees better coverage of the population than a complete randomized sampling.

For randomized treatment group allocation, we use stratification to minimize consecutive assignments. From previous allocation designs it is likely to observe 2 adjacent subjects are assigned into the same treatment group. We definitely do not want to see an uninterrupted succession of assigning subjects into one of the treatment groups. In another words, consecutive assignments into one group, AAAAA or BBBBB happened.

So we break the total number of test subjects into strata. Each stratum is called a block. Inside each block, we assign subjects into treatment groups by the random allocation designs we have described.

The following example uses the random allocation rule. Because the random allocation rule is an urn design, we put an equal number of A's and B's in the urn first. And the A's and B's are equally divided into blocks. Each A or B also has a random decimal number associated with it. Then we do the sorting within each block to randomize A's and B's.

# SAS Code

```
data block_design;

     M=50;
     sign=1;
     block_size=10;
* An alternating sequence of length of M;
     do i=1 to M;

* By multiply -1, we make the sign to alternate;
          sign=sign*(-1);
          * When the sign >0, assign A;
          if sign>0 then do;
               T='A';
               * Also give a uniform(0,1);
               x=ranuni(3267);
          end;

     * Otherwise, when the sign <0, assign B;
          else do;
               T='B';

               * Also give a uniform(0,1);
               x=ranuni(3267);
          end;

     * Assign block membership, 10 is block size;
     block_membership=int((i-1)/block_size+1);

          output;
     end;
run;

* Sort the uniform number within each block;
proc sort data=block_design;
     by block_membership x;
run;


proc print data=block_design;
     var t x block_membership;
run;
```

# Interpreting the Result

| Obs | T | x | block_ membership |
|---|---|---|---|
| 1 | A | 0.03686 | 1 |
| 2 | B | 0.22601 | 1 |
| 3 | B | 0.27272 | 1 |
| 4 | A | 0.27983 | 1 |
| 5 | A | 0.45537 | 1 |
| 6 | A | 0.47292 | 1 |
| 7 | A | 0.62910 | 1 |
| 8 | B | 0.67708 | 1 |
| 9 | B | 0.87562 | 1 |
| 10 | B | 0.91296 | 1 |
| 11 | A | 0.00006 | 2 |
| 12 | A | 0.00010 | 2 |
| 13 | B | 0.03921 | 2 |
| 14 | B | 0.09081 | 2 |
| 15 | A | 0.48499 | 2 |
| 16 | A | 0.49201 | 2 |
| 17 | B | 0.59694 | 2 |
| 18 | B | 0.62271 | 2 |
| 19 | B | 0.71377 | 2 |
| 20 | A | 0.93268 | 2 |
| 21 | B | 0.29385 | 3 |
| 22 | A | 0.36607 | 3 |
| 23 | B | 0.40366 | 3 |
| 24 | B | 0.46881 | 3 |
| 25 | A | 0.49955 | 3 |
| 26 | A | 0.57245 | 3 |
| 27 | A | 0.70268 | 3 |
| 28 | A | 0.72137 | 3 |
| 29 | B | 0.79629 | 3 |
| 30 | B | 0.99029 | 3 |
| 31 | B | 0.01062 | 4 |
| 32 | A | 0.02704 | 4 |
| 33 | A | 0.07373 | 4 |
| 34 | A | 0.14464 | 4 |
| 35 | B | 0.27481 | 4 |
| 36 | B | 0.42032 | 4 |
| 37 | A | 0.42612 | 4 |
| 38 | A | 0.43520 | 4 |
| 39 | B | 0.53563 | 4 |
| 40 | B | 0.55722 | 4 |
| 41 | B | 0.04078 | 5 |
| 42 | B | 0.14541 | 5 |
| 43 | A | 0.21559 | 5 |
| 44 | A | 0.27931 | 5 |
| 45 | A | 0.47464 | 5 |
| 46 | B | 0.47552 | 5 |
| 47 | B | 0.83820 | 5 |
| 48 | B | 0.86566 | 5 |
| 49 | A | 0.89966 | 5 |
| 50 | A | 0.96685 | 5 |

There are 50 test subjects and 5 blocks. Each block has 10 subjects in it, 5 of them need to be allocated to A, and other 5 of them need to be allocated to B.

We give each block 5 As and 5 Bs, each A or B has a random number. We randomize the permutation of A and B within each block by sorting the random number and the block membership.

# Regression with Treatment Effect and Interaction

## Problem Setup

There is a study on a certain type of neurotransmitter. 11 test subjects were randomly divided into 2 groups. The first group of 5 participants were administered the actual treatment and the second group of 6 test subjects received a placebo. Following the treatments the amount of the neurotransmitter were measured, and listed in the SAS code as variable y.

Before the treatments, another type of chemical was also measured from each test subject, and indicated by the SAS code as variable x. We call this variable a covariate because it may vary with present or absent of the treatment.

We assign numerical value 1 to the first group because they had treatment present, and we assigned 0 to the placebo group because they had no treatment. We also define another variable called interaction (int) which is product of treatment and covariate.

So far we can have 3 options to model the measured neurotransmitter amount by regression.
1. Model it only with the treatment.
2. Model it with treatment and covariate.
3. Model it with treatment, covariate, and the interaction between treatment and covariate.

## SAS Code

```
options nodate ls=72;
data one;
input y t x;
int = t*x;
datalines;
96      1       58
117     1       64
127     1       82
132     1       81
152     1       93
136     0       102
150     0       84
158     0       75
155     0       61
181     0       45
190     0       52

;
run;

proc reg data=one;
model y=t;
run;

proc reg data=one;
model y = t x;
run;

proc reg data=one;
model y = t x int;
run;
```

# Interpreting the Result

Parameter Estimates

| Variable | DF | Parameter Estimate | Standard Error | t Value | Pr > \|t\| |
|---|---|---|---|---|---|
| Intercept | 1 | 161.66667 | 8.29562 | 19.49 | <.0001 |
| t | 1 | -36.86667 | 12.30439 | -3.00 | 0.0150 |

This SAS result shows estimated parameters of the linear regression modeling on y only by using treatment t. The p-value of these 2 estimated parameters are very small. We can conclude the estimates are statistically significant and we should reject the null hypothesis that states the estimated parameters are zero.

The linear regression formula of modeling the average of y at a given t, only by t:
$$E(y|t) = b_0 + b_1 t$$
$b_0$ and $b_1$ are the intercept and slope of the linear regression model. We can plug in the estimated $b_0$ and $b_1$:
$$E(y|t) = 161.66667 - 36.86667 \times t$$

Linear regression model does not predict the exact value of the response variable y. Linear regression model estimates the average value of the response variable y at a given value of the predictor variable t.

Parameter Estimates

| Variable | DF | Parameter Estimate | Standard Error | t Value | Pr > \|t\| |
|----------|----|--------------------|----------------|---------|-----------|
| Intercept | 1 | 180.02219 | 27.63114 | 6.52 | 0.0002 |
| t | 1 | -35.35091 | 12.85458 | -2.75 | 0.0251 |
| x | 1 | -0.26285 | 0.37629 | -0.70 | 0.5046 |

This SAS result shows estimated parameters of the linear regression modeling on y by using treatment t and covariate x. And the p-value of the first 2 estimated parameters are quite small and we can conclude the first 2 estimates are statistically significant. On that basis,
we should reject the null hypothesis that states the first 2 estimated parameters are zero.

However, the p-value of parameter estimate for covariate x is above the usual level of significance 0.05. We failed to reject the null hypothesis of the parameter estimate for covariate x. It is plausible to say the parameter estimate of covariate x is zero.

The linear regression formula of modeling the average of y at a pair of given t and x:
$$E(y|t, x) = b_0 + b_1 t + b_2 x$$

We can plug in the estimated $b_0$, $b_1$ and $b_2$:
$$E(y|t, x) = 180.02219 - 35.35091 \times t - 0.26285 \times x$$

# Regression with Treatment Effect and Interaction

```
                    Parameter Estimates

                Parameter       Standard
Variable    DF   Estimate          Error    t Value    Pr > |t|
Intercept    1  221.10865       13.13918      16.83      <.0001
t            1 -200.25403       26.64903      -7.51      0.0001
x            1   -0.85120        0.18124      -4.70      0.0022
int          1    2.22614        0.35254       6.31      0.0004
```

This SAS result shows estimated parameters of the linear regression modeling on y by using treatment t, covariate x and interaction between treatment and covariate. The interaction is defined as $int = t \times x$.

The p-value of the 4 estimated parameters are quite small; we can conclude all the parameter estimates are statistically significant and we should reject the null hypothesis that states all the estimated parameters are zero.

The linear regression formula of modeling the average of y at a triple of given t, x and int:
$$E(y|t, x, int) = b_0 + b_1 t + b_2 x + b_3 int$$

We can plug in the estimated $b_0, b_1, b_2,$ and $b_3$:
$$E(y|t, x, int) = 221.108 - 200.254 \times t - 0.851 \times x + 2.226 \times x \times t$$

# Logistic Regression - Regression of Proportion

## Problem Setup

Can we use regression to model binary trial outcome? Trial outcomes will have both success and failure. Our mandate with this model is to determine the proportion of test subjects that can be counted as "successful". We will use regression to model this proportion.

The problem here is that a proportion as a decimal number has range between 0 and 1. However the value that can be modeled by regression is a real number with a range from negative infinite to positive infinite. We want to transform the proportion into a value which has the range of a real number.

Let's define $\pi$ as the proportion of success, and $(1-\pi)$ as the proportion of failure. So the $\log[\pi/(1-\pi)]$ as a value has the range of a real number. The ratio of $\pi/(1-\pi)$ is called odds of the proportion of success. We usually write log(odds) as logit($\pi$). Because logit($\pi$) is a real number, we can use regression to model it. We call it logistic regression.

The y column in the SAS code is the outcome of the trial with 1 representing success and 0 indicating failure. Two treatment groups are indicated by column t. The x column is for covariate measurement.

# SAS Code

```
data two;
input y t x;
int = t*x;
datalines;
0      1       53
0      1       67
0      1       66
1      1       91
1      1       96
1      1       190
0      1       106
0      1       85
1      1       154
1      1       113
1      1       135
1      1       110
0      1       56
0      0       95
0      0       76
0      0       78
1      0       79
1      0       64
1      0       52
0      0       161
1      0       127
0      0       72
0      0       125
0      0       132;
run;

proc logistic data=two des;
model y=t;
run;

proc logistic data=two des;
model y=t x;
run;

proc logistic data=two des;
model y=t x int;
run;
```

# Interpreting the Result

```
                        The LOGISTIC Procedure
              Analysis of Maximum Likelihood Estimates

                                Standard        Wald
   Parameter    DF    Estimate     Error    Chi-Square    Pr > ChiSq
   Intercept    1     -0.5596     0.6268      0.7971        0.3720
   t            1      0.7137     0.8381      0.7253        0.3944
```

We can see that the p-values of the 2 parameter estimates are a lot bigger than the usual level of significance 0.05. So it is plausible to say the parameter estimates are zero.

Let's define p to be proportion of success on a treatment group, also define $\text{logit}(p) = \log(\frac{p}{1-p})$. We use linear regression to model $\text{logit}(p)$. We call it logistic regression:

$$\text{logit}(p) = b_0 + b_1 t$$

Equivalently:

$$p = \frac{e^{b_0 + b_1 t}}{1 + e^{b_0 + b_1 t}}$$

Plug in the estimates above:

$$\text{logit}(p) = -0.56 + 0.71 t$$

$$p = \frac{e^{-0.56 + 0.71 t}}{1 + e^{-0.56 + 0.71 t}}$$

```
                        The LOGISTIC Procedure
            Analysis of Maximum Likelihood Estimates

                              Standard          Wald
Parameter     DF    Estimate     Error    Chi-Square    Pr > ChiSq
Intercept     1     -2.1440     1.4287       2.2521        0.1334
t             1      0.6884     0.8696       0.6268        0.4285
x             1      0.0161     0.0127       1.6001        0.2059
```

From our previous logistic regression definition, we can model the logit(p) with these 3 estimated parameters which consider treatment t as well as covariate x.

$$\text{logit}(p) = b_0 + b_1 t + b_2 x$$

Equivalently:

$$p = \frac{e^{b_0 + b_1 t + b_2 x}}{1 + e^{b_0 + b_1 t + b_2 x}}$$

Plug in the estimates above:

$$\text{logit}(p) = -2.14 + 0.69t + 0.016x$$

$$p = \frac{e^{-2.14 + 0.69t + 0.016x}}{1 + e^{-2.14 + 0.69t + 0.016x}}$$

```
                   The LOGISTIC Procedure
        Analysis of Maximum Likelihood Estimates

                            Standard         Wald
  Parameter    DF   Estimate    Error   Chi-Square   Pr > ChiSq
  Intercept    1      2.0145    2.2451     0.8051       0.3696
  t            1    -12.4961    6.6764     3.5032       0.0613
  x            1     -0.0280    0.0246     1.2975       0.2547
  int          1      0.1406    0.0705     3.9761       0.0461
```

From our previous logistic regression definition, we can model the logit(p) with these 4 estimated parameters which consider treatment t, covariate x as well as interaction between x and t.

We define $int = x \times t$

$$\text{logit(p)} = b_0 + b_1 t + b_2 x + b_3 int$$

Equivalently:

$$p = \frac{e^{b_0 + b_1 t + b_2 x + b_3 int}}{1 + e^{b_0 + b_1 t + b_2 x + b_3 int}}$$

Plug in the estimates above:

$$\text{logit(p)} = 2.01 - 12.5t - 0.028x + 0.14 int$$

$$p = \frac{e^{2.01 - 12.5t - 0.028x + 0.14 int}}{1 + e^{2.01 - 12.5t - 0.028x + 0.14 int}}$$

# Proportional Hazards Model

# Regression of Survival Data

## Problem Setup

Survival trials are medical trials with binary outcomes. We can use logistic regression as well to model survival. We have to notice that a higher percentage of subjects are censored, meaning more people survived the trial. Just because one group produced these results, it does not necessarily indicate this treatment is better than the other one.

The other treatment group, for instance, may have a longer average survival time during the trial and most of its subjects may have entered the trial very early. And the trial can be just too long to let subjects last until the trial finishes.

There are 4 variables that need to be considered in order to model the survival data under the logistic regression umbrella. As per the data list in the SAS code, the first variable indicates each test subject's survival time during the trial. The second variable indicates whether the test subject is censored. If the test subject is still alive when the trial finishes, we assign 1 to the subject as he gets censored. The third variable is treatment group indication specifies we have 2 treatment groups, treatment 1 and treatment 2. The fourth variable is a covariate measurement, because we may have interest in this covariate and its interaction with the treatment effect.

# SAS Code

```
options ls=72;
data one;
input time censor group covar;
int=group*covar;
cards;
6        0       1       23
7        0       1       18
6        0       1       24
8        1       1       19
8        0       1       16
16       0       1       16
20       1       1       18
21       0       1       16
25       0       1       24
25       1       1       27
25       0       1       18
26       0       1       17
26       0       1       21
27       0       1       14
3        1       2       24
6        0       2       21
6        0       2       20
6        0       2       13
5        1       2       26
8        0       2       23
9        0       2       22
6        0       2       15
8        0       2       31
8        0       2       26
16       0       2       22
21       0       2       20
19       1       2       25
19       0       2       24
21       0       2       26
22       0       2       23
28       0       2       25
;

proc lifetest;
```

```
      time time*censor(1);
      strata group;
      title 'Comparison of Survival Curves';
      label group='treatment group'
      time='death time';
run;

*modeled only by treatment group;

proc phreg covout outest=out data=one;
      model time*censor(1)=group/rl;
run;

*modeled by treatment group and covariate;
proc phreg covout outest=out1 data=one;
      model time*censor(1)=group covar/rl;
run;

*model by treatment group and covariate and
interaction between treatment and covariate;
proc phreg covout outest=out2 data=one;
      model time*censor(1)=group covar int/rl;
run;
```

# Interpreting the Result

```
           Test of Equality over Strata

                                      Pr >
Test        Chi-Square     DF     Chi-Square
Log-Rank      1.9219        1       0.1656
Wilcoxon      2.3458        1       0.1256
-2Log(LR)     0.9684        1       0.3251
        Comparison of Survival Curves
```

First of all, from the Log-Rank p-value=0.1656 of the life test procedure, we can see the two survival curves are not significantly different. If we set the level of significance to 0.05, we are able to conclude that the survival time from these 2 treatment groups are not significantly different.

Analysis of Maximum Likelihood Estimates

| Parameter | DF | Parameter Estimate | Standard Error | Chi-Square | Pr > ChiSq |
|-----------|-----|--------|--------|------------|-----------|
| group | 1 | 0.54208 | 0.42509 | 1.6261 | 0.2022 |

Let t be the survival time. Let T be the treatment group indicator, T=1 for treatment 1, T=2 for treatment 2. We need to specify a underlying hazard function $S_0(t)$.

So we have the Cox Proportional Hazards Model which considers only the treatment group variable:

$$S(t) = S_0(t)e^{b_1 T}$$

We can now plug in the parameter estimate from the SAS output:

$$S(t) = S_0(t)e^{0.54208 T}$$

```
                  The PHREG Procedure
          Analysis of Maximum Likelihood Estimates

              Parameter      Standard
Parameter    DF   Estimate      Error    Chi-Square    Pr > ChiSq
group         1    0.94525    0.53283      3.1471          0.0761
covar         1   -0.07974    0.06026      1.7511          0.1857
```

Let t be the survival time. Let T be the treatment group indicator, T=1 for treatment 1, T=2 for treatment 2. Let x be the covariate variable. We need to specify a underlying hazard function $S_0(t)$.

So we have the Cox Proportional Hazards Model considers the treatment group and covariate variables:

$$S(t) = S_0(t)e^{b_1 T + b_2 x}$$

We can now plug in the parameter estimates from the SAS output:

$$S(t) = S_0(t)e^{0.94525 T - 0.07974 x}$$

```
                      The PHREG Procedure
           Analysis of Maximum Likelihood Estimates

                    Parameter      Standard
Parameter    DF     Estimate        Error     Chi-Square    Pr > ChiSq
group        1       3.99108       2.41946      2.7211        0.0990
covar        1       0.14039       0.17783      0.6233        0.4298
int          1      -0.14778       0.11434      1.6704        0.1962
```

Let t be the survival time. Let T be the treatment group indicator, T=1 for treatment 1, T=2 for treatment 2. Let x be the covariate variable. Also define the interaction variable as $int = T \times x$. We need to specify a underlying hazard function $S_0(t)$.

So, we have the Cox Proportional Hazards Model which considers the treatment group and covariate variables as well the interaction:

$$S(t) = S_0(t)e^{b_1 T + b_2 x + b_3 int}$$

We can now plug in the parameter estimates from the SAS output:

$$S(t) = S_0(t)e^{3.99018 T + 0.14039 x - 0.14778 int}$$

www.ingramcontent.com/pod-product-compliance
Lightning Source LLC
LaVergne TN
LVHW052306060326
832902LV00021B/3729